Strange
Sea Creatures

ERICH **HOYT**

FIREFLY BOOKS

A FIREFLY BOOK

Published by Firefly Books Ltd. 2020
Copyright © 2020 Firefly Books Ltd.
Text copyright © 2020 Erich Hoyt

First printing

Library of Congress Control Number: 2020932460

Library and Archives Canada Cataloguing in Publication
Title: Strange sea creatures / Erich Hoyt.
Names: Hoyt, Erich, author.
Description: Includes index.
Identifiers: Canadiana 20200190334 |
    ISBN 9780228102977 (hardcover)
Classification: LCC QL121 .H69 2020 |
    DDC 591.77—dc23

Published in the United States by
Firefly Books (U.S.) Inc.
P.O. Box 1338, Ellicott Station
Buffalo, New York 14205

Published in Canada by
Firefly Books Ltd.
50 Staples Avenue, Unit 1
Richmond Hill, Ontario L4B 0A7

Cover and interior design: Stacey Cho
Front cover: Alexander Semenov
Back cover: David Shale (top), Susan Mears (bottom)
Back cover flap: Max Hoyt (top), Susan Mears (bottom)

Printed in China

Canada ✦ We acknowledge the financial support of the Government of Canada.

For my sisters, Gerard and Victoria Hoyt, Karres Cvetkovich and Christina Innes, and daughters Jasmine and Magdalen Hoyt, with love and admiration.

AUTHOR'S NOTE

Thank you to my collaborators at Firefly Books including Michael Worek for editing the book, Steve Cameron for his help in shaping the book and negotiating the photographs, Stacey Cho for her design and layout, and Lionel Koffler for his valuable ideas and making the book possible in the first place. Christi Linardich and David Shale read the entire text and supplied welcome corrections and comments. In addition, Linda Ianniello, Susan Mears and Alexander Semenov read the sections with their photographs and gave corrections. I am especially grateful to David, Linda, Susan and Alex, and all the photographers listed below, as well as Rachelle Morris at Nature Picture Library, for providing photographs and for various email exchanges to answer questions and provide further information.

PHOTO CREDITS

Alexander Semenov: 34, 35, 42–43, 44, 45, 46, 47, 63, 64, 65, 78, 80, 81, 82, 84, 85, 86, 87, 101

Linda Ianniello: 10, 11, 12, 13, 14, 15, 16, 17, 18, 19, 20, 21, 22, 29, 30, 37, 38, 39, 40

Ocean Exploration Trust and NOAA ONMS: 106

Susan Mears: 23, 24, 25, 26, 27, 28, 36, 41

Nature Public Library
Alex Mustard: 72, 75
David Shale: 50, 51, 52, 53, 54, 55, 56, 57, 58, 62, 66, 68, 69, 88, 89, 91, 92, 93, 94, 95, 96, 97, 102, 103, 104–105
Doc White: 48–49
Doug Perrine: 31, 32–33
Franco Banfi: 70–71, 98
Georgette Douwma: 77
Shane Gross: 6–7, 8–9
Solvin Zankl: 59, 60, 61, 67, 99

CONTENTS

INTRODUCTION

Most of the animals in this book could fit easily into your open hand. Many are less than an inch (2.5 cm) long; some are so small you would need a microscope to see them. The images in this book give you a chance to experience an alien world by gazing at these deep-sea creatures eye-to-eye or, in the case of no-eyed sea cucumbers, mouth to rear end.

If you were as small as some of the larval animals in this book, you would not want to be in the ocean on a dark night unless you had some way to protect yourself. It would help if you had stinging tentacles filled with poison to fight predators. Or if you were transparent, then predators could see through you. Or if you had flashing lights from bioluminescence, you could fool your predators into thinking you were much larger and more dangerous than you really are. Then, other creatures would stay away.

The predators you need to worry most about are those that are only a little bit bigger than you. The big toothy sharks, 20 feet (6 m) long, the giant squid at 43 feet (13 m) and the colossal squid at up to 46 feet (14 m) long, are not dangerous to you. When you are only an inch (2.5 cm) or so long, your size, in fact, keeps you safe from most of the big animal species who cannot see you and couldn't be bothered to chase you.

Yet, as it happens in the sea, eventually you would probably encounter the gobbling mouth of a ravenous squid or a fish a little bit larger than you are. Then you would instantly become part of the food chain — a meal for another creature.

Many of the animals in this book are naturally small; others are small because they are larvae, the young form of a species that gets larger as it matures. Some of the animals look terrifying to us with their big-eyed stare, their flashing long sharp teeth, their barbed tentacles or their spiky backs. Others look as tame as a garden plant. The sea anemones and the stalked jellyfish look more like flowers. Others have names that make you think they're a vegetable or fruit — sea cucumbers, sea apples and sea grapes. The animals that live on the bottom of the ocean have no real face, no eyes, no ears. Only a hole at either end, a mouth and an anus. They spend most of their lives sucking and sifting sand and sediment on the bottom and then ejecting most of it, hoping to get enough nutrients to survive.

This book celebrates the work of photographers who have explored every ocean using special techniques to uncover the mysteries of the darkest depths. Many are divers, too, and all are experienced members of sea-going expeditions.

- David Shale worked on the BBC Blue Planet series and his photos range from offshore sites in the Gulf of Maine and hydrothermal vents along the Mid-Atlantic Ridge in the North Atlantic Ocean to the Southwest Indian Ridge in the Indian Ocean.

- Linda Ianniello and Susan Mears have specialized in blackwater dives at the edge of the Gulf Stream off southern Florida and have ventured to the remote islands of the Philippines and Indonesia. Blackwater refers to the nighttime surface layer conditions which create nocturnal feeding frenzies of deep to midwater species, the larvae of which migrate close to the surface.

- Alex Semenov has photographed his subjects in the Russian Arctic, mainly in the White Sea of northwest Russia, but also in the Russian Far East waters of the Okhotsk Sea.

These and other photographers are biologists themselves, or work closely with marine biologists, on the frontiers of human knowledge. Some of the species they have captured are newly identified and named — partly the result of the Census of Marine Life (2000 to 2010) and follow-up research. Others have been discovered but the work to identify them remains incomplete. In addition, some of the species shown here, whether they have a Latin scientific name or not, have no common name.

There are many more species in the sea that have yet to be discovered and named. My other books *Creatures of the Deep* and *Weird Sea Creatures* featured mainly middle to the deepest deep-sea creatures. This volume explores the underwater world at various depths of the ocean. An illustrated book covering the world oceans' animals may be possible one day but to be complete, even with only one page for each new species, it would have to be many hundreds of thousands, perhaps a million or more, pages long.

The book you have in front of you is only a sample of what we know now. Enjoy.

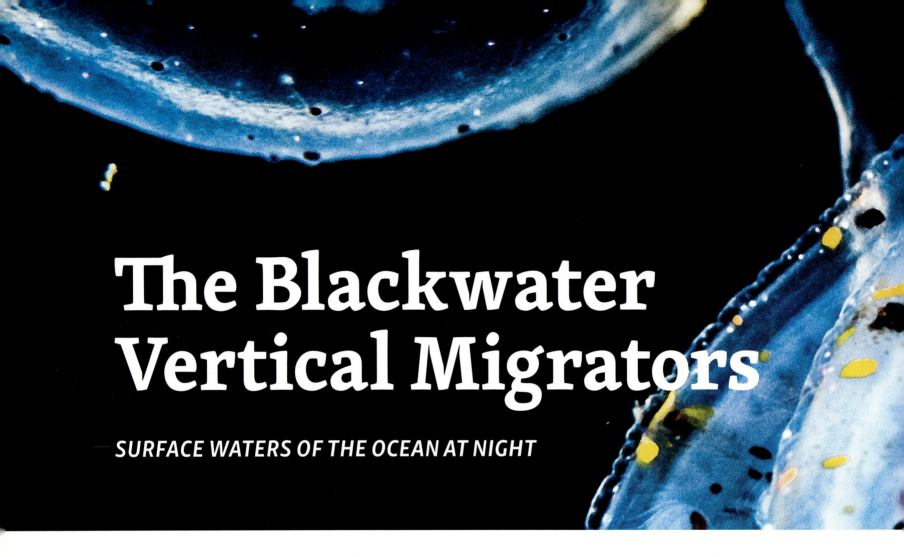

The Blackwater Vertical Migrators

SURFACE WATERS OF THE OCEAN AT NIGHT

Once a day, as the Earth turns away from the sun and the sea's surface on one side of the globe turns dark, millions of hungry ocean creatures swim up the water column. In waters just beneath the surface under the cover of night, they gobble up the rich nutrients and tiny plant and animal plankton. At the same time, they try to avoid being eaten themselves. Everywhere are hungry mouths, predators trying to survive.

Many of these vertical migrators are tiny planktonic fish, crustaceans, squid, octopus and other creatures at the early stages of their development. Some resemble a tiny version of their future selves but many look completely different, and indeed may be unrecognizable. Each of these creatures is otherworldly in its own way. They have evolved bizarre, often transparent body shapes and special eyes, mouths, teeth, predatory and defensive behavior to flourish in their daily passage up-and-down their watery world.

Dedicated blackwater photographers Linda Ianniello and Susan Mears have made hundreds of

night time dives to take close-up underwater shots
capturing the most intimate behavior of these
hard to identify creatures that live at the western
edge of the Gulf Stream off Florida as well as in the
waters around the Philippines and Indonesia.

Wunderpus Octopus
Wunderpus photogenicus

The wunderpus octopus was discovered in the 1980s, but it was only recently described in detail. Each octopus has a unique pattern of white spots on its head which allows researchers to identify and follow the lives of specific individuals. The one shown here, a 4-inch (10 cm) long larval form, was photographed at night where it had come to within 10 feet (3 m) of the surface off Batangas Bay, near Anilao, Philippines. If it survives, this larva will grow to 10 inches (23 cm) long with a 16 inch (40 cm) wingspan. They burrow into the soft sediment on the bottom during the day and hunt crustaceans and fish by night. Like some other squid and octopuses, the wunderpus appears to be able to drop an arm when attacked and thus escape. They can regrow the arm later.

Pelagic Tunicate
order Doliolida, unidentified species

Found at a depth of 30 feet (9 m) below the surface near the western edge of the Gulf Stream off southeast Florida, this free-swimming 1⁹⁄₁₆ inch (4 cm) long filter-feeding tunicate pumps water through its body to push itself forward. As the water passes through, small particles and plankton on which the animal feeds are strained from the water by the gill slits. The flowing water thus efficiently provides both movement and food.

Atlantic Flyingfish

Cheilopogon melanurus

This 1 inch (2.5 cm) long flyingfish was found only 5 feet (1.5 m) below the surface of the ocean. When fully mature, it grows to more than 12 inches (32 cm) long. Atlantic flyingfish are capable of leaping out of the water and gliding for long distances above the surface.

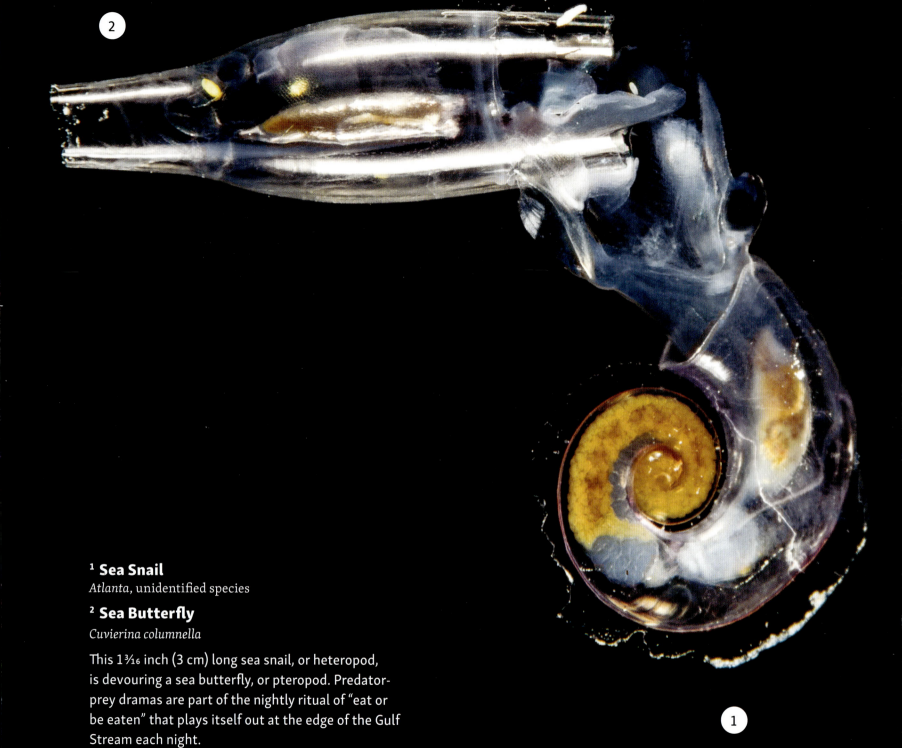

¹ Sea Snail

Atlanta, unidentified species

² Sea Butterfly

Cuvierina columnella

This 1³⁄₁₆ inch (3 cm) long sea snail, or heteropod, is devouring a sea butterfly, or pteropod. Predator-prey dramas are part of the nightly ritual of "eat or be eaten" that plays itself out at the edge of the Gulf Stream each night.

Hydrozoan
Oceania armata

Found in the tropical to warm temperate waters of the Atlantic, Indian and Pacific oceans, this hydrozoan, shown here in the medusa, or free-swimming, jellyfish stage, was photographed swimming at a depth of about 40 feet (12 m) below the surface of the sea in the bay at Ambon, Indonesia. The mantle, or bell-shaped dome of this medusa, the main part of its body, is about ⅜ inch (1 cm) across. In part of its life cycle it also takes on the body form of a polyp living on the surface of rocks or on the ocean bottom.

Cleaner Shrimp (left)
Hippolytidae or Lysmatidae, unidentified species

This planktonic larva, less than an inch (2 cm) long, comes to within 30 feet (9 m) of the surface at night. When it becomes an adult, it settles on the ocean bottom which, in this case, is up to 600 feet (180 m) below the surface of the ocean off southeast Florida at the edge of the Gulf Stream. Exact identification is difficult as few scientists currently work on shrimp larvae.

Eel
Leptocephalus stage, unidentified species

Any attempt at elegance in the eel world is found in the larval or Leptocephalus (meaning "slim head") stage. The 2⅜ inch (6 cm) long body looks like a flat, transparent ribbon being dragged behind a snake-like head with a mouth full of teeth. At this stage, it is difficult to identify larvae to the species level. When these eels grow into adults, they settle permanently on the bottom but, as ravenously hungry larvae, they need to make vertical migrations to find enough food to grow and develop.

Spotfin Lionfish
Pterois antennata

Photographed 50 feet (15 m) below the surface of the ocean in a bay at Ambon, Indonesia, this ¾ inch (2 cm) long larva has already developed features of a mature lionfish. As an adult, an 8 inch (20 cm) long spotfin lionfish will continue its night time hunting habits. To trap the larger crabs and shrimp it preys on for food, a lionfish fans out its fins as far as possible. Its dorsal and pectoral fins are formed of long, free-moving white spines, each one connected to a venom gland that stuns prey on contact. By day, spotfin lionfish take shelter in rocky or coral crevasses, where they sometimes rest upside down, head facing toward the bottom.

Paper Argonaut or Paper Nautilus
Argonautidae – Argonauta, unidentified species

The argonauts are a group of pelagic octopuses found in deep tropical and subtropical waters worldwide. This ¾ inch (2 cm) long female paper nautilus was photographed 60 feet (18 m) below the surface near Anilao, Philippines. The water there is approximately 200 feet (61 m) deep. The term "paper nautilus" refers to the paper-thin egg case that the females secrete.

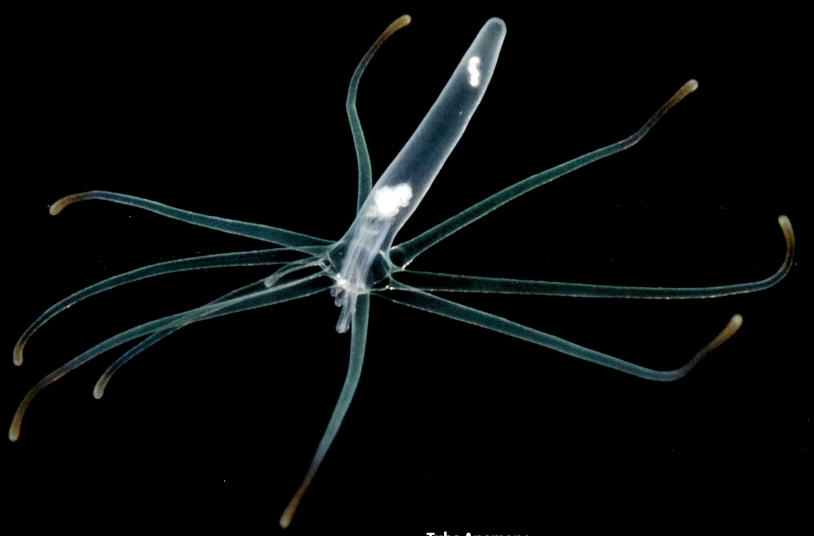

Tube Anemone
subclass Ceriantharia, unidentified species

This larval stage of a tube anemone is 1 inch (2.5 cm) wide. It was found at a depth of 25 feet (7.6 m) below the surface along the western edge of the Gulf Stream off southeast Florida. Tube-dwelling anemones, or ceriantharians, look similar to sea anemones but belong to a different subclass. Once mature in its flowerlike form, it can be identified as a species. The solitary tube anemone settles on the bottom, buries itself in the sand leaving only its tentacles waving back and forth to catch particles of food.

Sea Butterfly
Cavolinia tridentata

This ¾ inch (2 cm) long sea butterfly glides through the night waters on what look like leafy wings. Sea butterflies are so named because of the way they swim and float through the water with a motion much like a butterfly in flight. In fact, these creatures are mollusks, properly sea snails, and are sometimes referred to as sea slugs. Sea butterflies feed by secreting a spherical web of mucus many times the size of their body, to which mainly phytoplankton and small zooplankton stick. This web also serves as a buoyancy mechanism to stop the animal sinking when its wings stop beating. The web can be set in about 5 seconds and retracted in less than 20 seconds.

Lined Seahorse
Hippocampus erectus

Juvenile seahorses swim through the forests of *Sargassum* seaweed in the Gulf Stream off Florida, attaching themselves to the seaweed to ride the currents. This 1¾ inch (4.5 cm) young seahorse won't undertake vertical migrations but as a poor swimmer will spend day and night in the *Sargassum*.

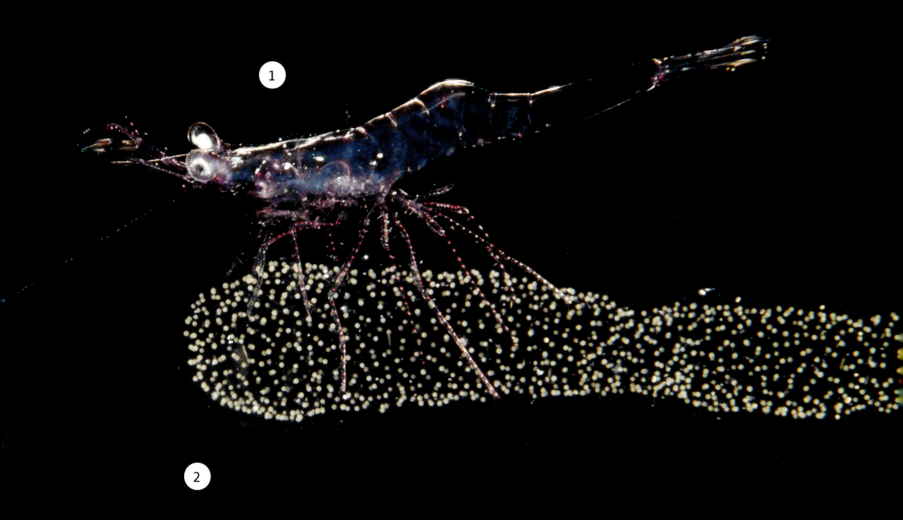

¹ Shrimp
unidentified species

² Radiolarian Colony
unidentified species

Less than 1 inch (2 cm) long, this unidentified larval shrimp catches a daring ride on a radiolarian colony. Radiolarian colonies consist of hundreds of single-cell animals embedded in a jellylike substance. They drift with the surface currents, catching and eating other small plankton.

Tonguefish (left)
Symphurus, unidentified species

The tonguefish is a flatfish in the family Cynoglossidae with a small mouth and both eyes on the left side of the head. The 1 inch (2.5 cm) long larva shown here has a protruding abdomen that allows a clear view of the gut tube inside. Some tonguefish have long dorsal fin filaments that give them a festive look. Tonguefish larvae move up to the surface waters to feed at night but as adults they settle on open sand or mud bottoms, sometimes in deep seas.

Acorn Worm
class Enteropneusta, unidentified species

This acorn worm is more attractive in its larval, or tornaria, stage than in its adult form. This photograph of a ¼ inch (0.6 cm) long larva was taken 30 feet (9 m) below the surface. As adults, acorn worms settle on the bottom and feed on the sediment of whatever ends up on the bottom of the sea.

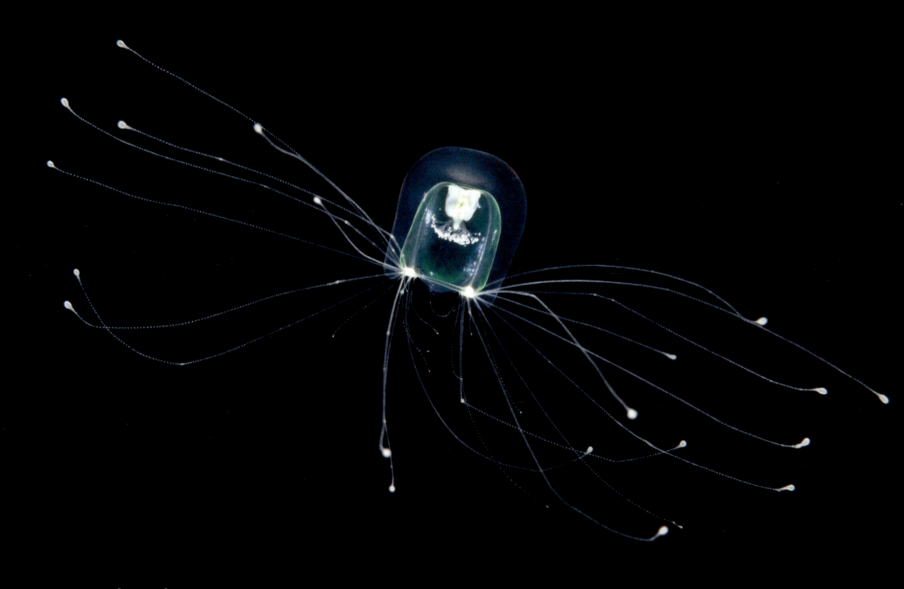

Hydromedusa
Bougainvillia, unidentified species

Named after the colorful bougainvillea flower, this hydromedusa, a jellyfish produced from a hydroid by budding, has a bell-shaped dome only 1 9/16 inches (4 cm) across, with glowing tentacle tips. It was found at a depth of 30 feet (9 m) below the surface at the edge of the Gulf Stream.

Spotfin Flounder
Cyclopsetta fimbriata

The 1 inch (2.5 cm) long flamboyant spotfin flounder larva, is frequently encountered on night dives at the western edge of the Gulf Stream at depths of about 25 feet (7.6 m) below the surface. When mature, it grows to more than 1 foot (33 cm) long. It will then make one final migration down to the soft bottom at a depth of up to 755 feet (230 m) below the surface. The eyes on the left side of its head looking up, are well camouflaged but they miss nothing.

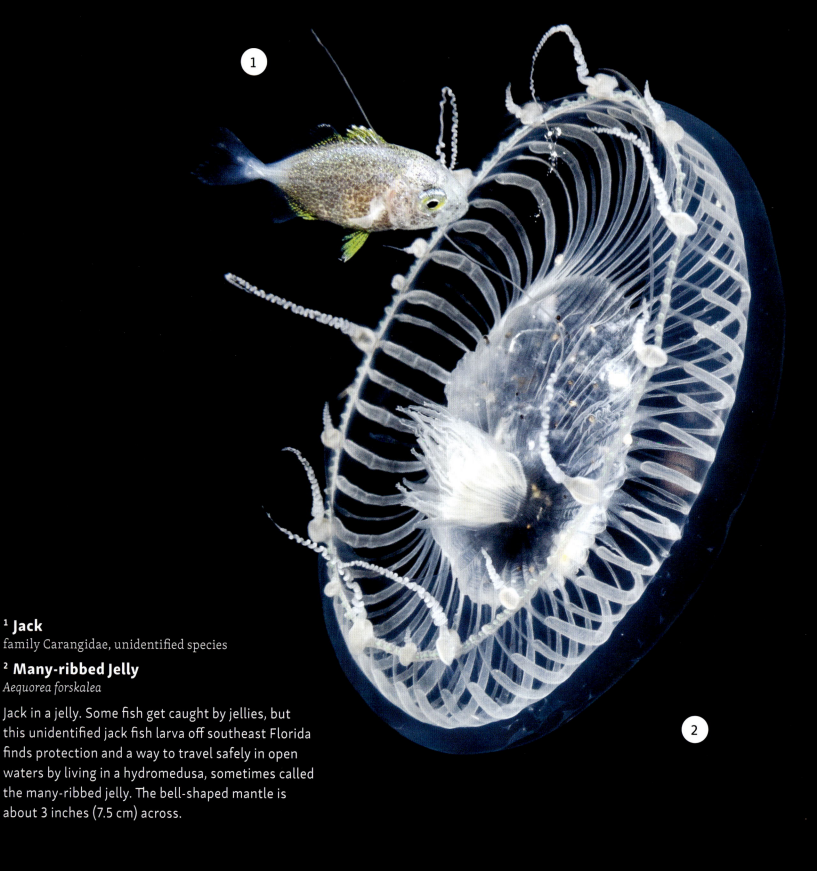

¹ Jack
family Carangidae, unidentified species

² Many-ribbed Jelly
Aequorea forskalea

Jack in a jelly. Some fish get caught by jellies, but this unidentified jack fish larva off southeast Florida finds protection and a way to travel safely in open waters by living in a hydromedusa, sometimes called the many-ribbed jelly. The bell-shaped mantle is about 3 inches (7.5 cm) across.

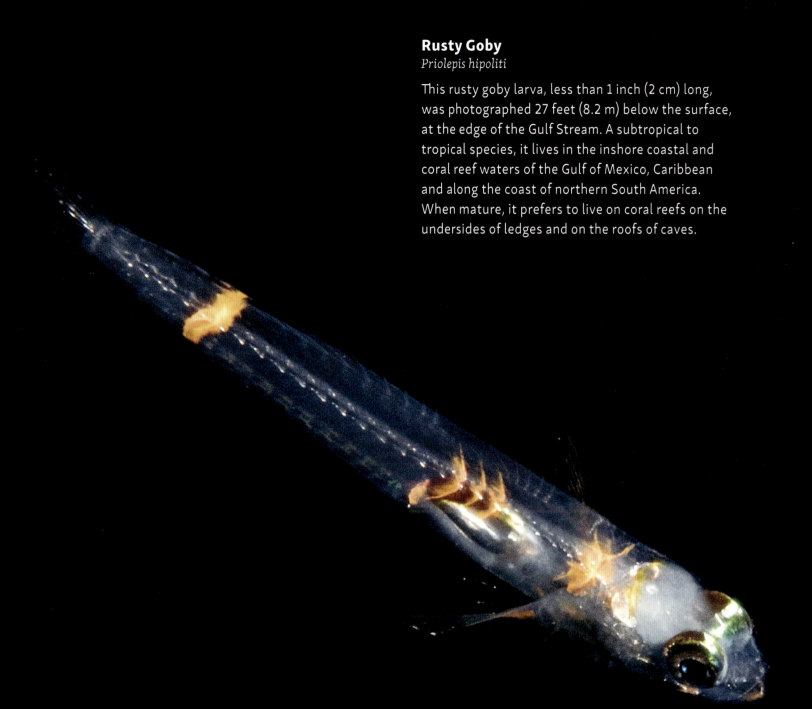

Rusty Goby
Priolepis hipoliti

This rusty goby larva, less than 1 inch (2 cm) long, was photographed 27 feet (8.2 m) below the surface, at the edge of the Gulf Stream. A subtropical to tropical species, it lives in the inshore coastal and coral reef waters of the Gulf of Mexico, Caribbean and along the coast of northern South America. When mature, it prefers to live on coral reefs on the undersides of ledges and on the roofs of caves.

Snaketooth Swallower
Kali, unidentified species

Only 2 inches (5 cm) long, the snaketooth swallower, in its larval form, has journeyed all the way to the topmost layer of the ocean to feed. Swallowers are deep-sea creatures common to worldwide tropical and subtropical waters. Once mature, living full-time in the deep, they reach up to 10 inches (25 cm) in length and have the ability to swallow fish larger than they are — up to twice the length and ten times the mass for one swallower species. If the fish cannot be digested before decomposition sets in, however, the resulting release of gasses forces the swallower to the surface where, as an adult, it cannot survive.

Shrimp with Hairy Legs
unidentified species

Shrimp larvae are notoriously hard to identify to the species level but this one may belong to the Sergestidae family of prawn species. The photographer who took this picture called this 1 inch (2.5 cm) long creature with its eyes on stalks and its bottle brush legs "shrimp larva with hairy legs." It was found at night in the waters off the Gulf Stream at a depth of 25 feet (7.6 m) below the surface.

Sharpear Enope Squid
Ancistrocheirus lesueurii

A resident of the tropical and subtropical oceans, this squid spends most of its time deep in the dark mesopelagic region, 656 to 3,280 feet (200 to 1,000 m) below the surface, trying to avoid hungry sperm whales. In the larval stage, it migrates vertically to feed on plankton found at a depth of about at 30 feet (9 m) below the surface. This transparent larva, about 1.2 inches (30 mm) long, shows signs of the adult it will become as it acquires photophores all over its body and becomes heavily pigmented. The adult mantle, the outside covering of the squid's body which protects its internal organs, measures 10 inches (25 cm) long.

¹ **Slipper Lobster**
unidentified species

² **Mauve Stinger Sea Jelly or Purple Jellyfish**
Pelagia noctiluca

This slipper lobster in the larval, or phyllosoma, stage was photographed at night in the surface waters off Hawaii. It has a leg span of 1³⁄₁₆ to 1⁹⁄₁₆ inch (3 to 4 cm). In this photograph, the lobster is riding a mauve stinger sea jelly, using the sea jelly as a defensive weapon, spinning it around to keep the stinging tentacles in motion to keep predators away. The slipper lobster may also eat the jellyfish. The diameter of the mantle, or bell-shaped dome, of mauve stinger sea jellies ranges from 1³⁄₁₆ to 4³⁄₄ inches (3 to 12 cm).

³ **Blue Dragon or Blue Sea Slug**

¹ Sea Slug
probably *Glaucilla marginata*

² Portuguese Man O' War
Physalia utriculus

³ Blue Dragon or Blue Sea Slug
Glaucus atlanticus

At the surface, off Kona, Hawaii, two sea slugs, one over ¾ inch (2 cm) long and the other ¼ inch (.75 cm) long, ride a Portuguese man o' war that is 1⁹⁄₁₆ inches (4 cm) long. The sea slugs are also feeding on the man o' war as they ride along. They store the venomous stinging cells, or nematocysts, from the man o' war and use them for their own defence.

Lion's Mane Jellyfish (left)
Cyanea capillata

Found at a depth of 105 feet (32 m) below the surface in a bay along the eastern Kamchatka coast, this giant-size jellyfish has a diameter of 3½ feet (1 m) with thousands of tentacles, 25 to 30 feet (7.6 to 9 m) long, extending down from the body. In Arctic circumpolar waters, however, the tentacles grow much longer. In 1865, oceanographer-biologist Alexander Agassiz found and measured a record specimen off the coast of Massachusetts with the bell-shaped mantle having a diameter of 7 feet (2.1 m) and tentacles that were 112 feet (34 m) long. The lion's mane mainly lives near the surface, using its sticky, stinging tentacles to capture, pull in, and eat zooplankton, small fish, ctenophores and moon jellies. Leatherback sea turtles, in turn, feed on large quantities of the lion's mane jellyfish during the summer season off Eastern Canada.

¹ Clapper Hydromedusa
Sarsia tubulosa

² Lion's Mane Jellyfish
Cyanea capillata

This clapper hydromedusa, less than ⅝ of an inch (1.5 cm) long, has been caught by a tentacle from a lion's mane jellyfish. The tiny gelatinous zooplankton like this hydromedusa, along with other jellyfishes and larvae, comprise most of the diet of the lion's mane.

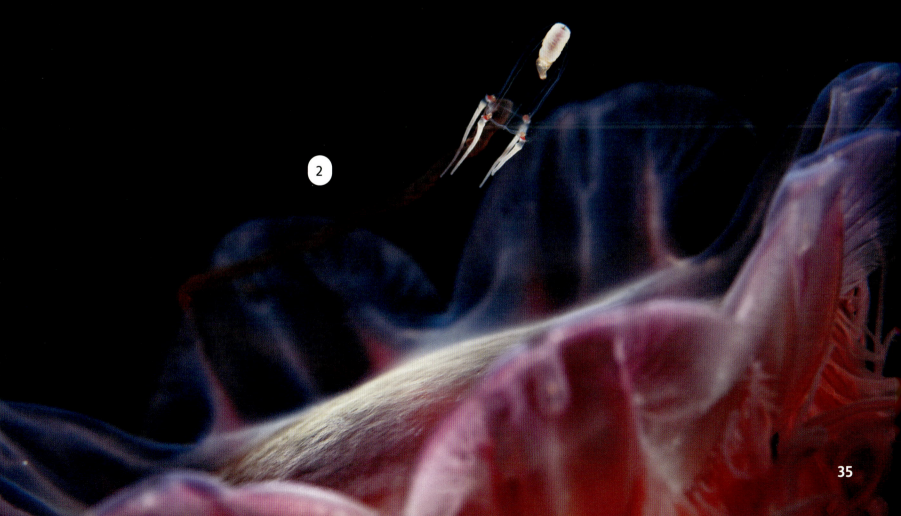

Spotfin Flounder
Cyclopsetta fimbriata

This 1 inch (2.5 cm) long larval flounder was photographed 30 feet (9.1 m) below the surface, 5 to 6 miles (8 to 9.7 km) off the southeast Florida coast. Larval flounders have eyes on either side of the head, but as they grow older the eyes migrate to one side so that the flounder can see while lying on its side at the bottom. One eye of this young flounder has not yet migrated to the other side.

Tonguefish (right)
Paraplagusia species

This developing young tonguefish, 2 inches (5 cm) long, was photographed in a bay at Ambon, Indonesia, at a depth of 25 feet (7.6 m) below the surface. The eyes are normally on the left side of the head, but this one has only one eye showing. The other eye should have migrated to the left hand side but hasn't yet. The red organism sprouting from its back is a parasite, the hydroid stage of a hydromedusa.

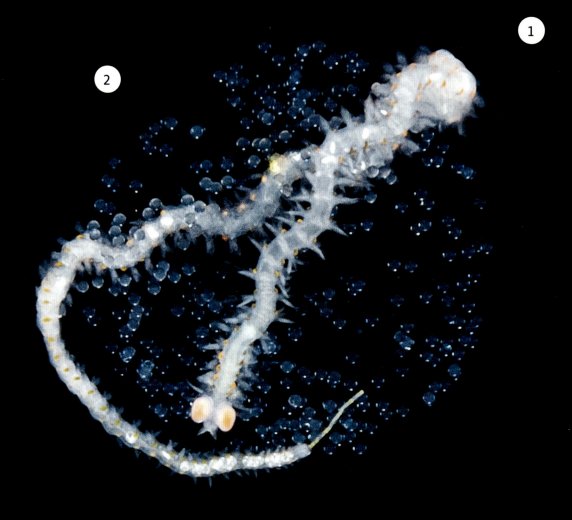

¹ **Polychaete Worm**
Tomopteris species

² **Sea Angel Eggs**
unidentified species

At a depth of 40 feet (12.2 m) below the surface at the edge of the Gulf Stream off southeast Florida, the nightly predator-prey drama is in process. A 3 inch (7.5 cm) long polychaete worm picks off what are thought to be sea angel eggs from the egg ball, devouring them one by one.

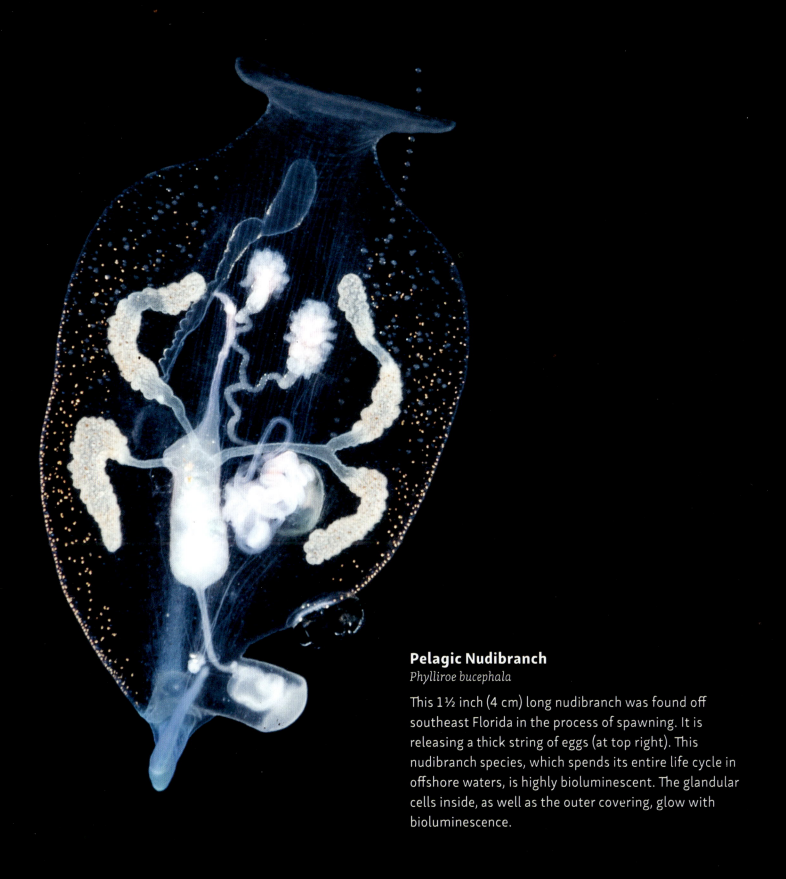

Pelagic Nudibranch
Phylliroe bucephala

This 1 ½ inch (4 cm) long nudibranch was found off southeast Florida in the process of spawning. It is releasing a thick string of eggs (at top right). This nudibranch species, which spends its entire life cycle in offshore waters, is highly bioluminescent. The glandular cells inside, as well as the outer covering, glow with bioluminescence.

Dragonfish
Bathophilus species

This 1½ inch (4 cm) long larval dragonfish was photographed 50 feet (15.2 m) below the surface at the western edge of the Gulf Stream off Florida. The greatly enlarged intestine is a characteristic of *Bathophilus* larvae. While in the larval form, it is difficult to identify the precise dragonfish species unless it is collected for DNA analysis.

Arrow Squid
Doryteuthis plei

Found at a depth of 20 feet (6 m) below the surface off southeast Florida, this juvenile arrow squid, measuring only 1¼ inches (3 cm) long, already resembles the mature individual. It may achieve a mantle length of up to 13 inches (33 cm) long. As adults, these squid live in great numbers along North and South Atlantic coastal waters where they feed on fish and small crustaceans. In turn, they are preyed upon by killer whales, various dolphins, tuna and sharks. Sometimes called the slender inshore squid, they are caught in large numbers by local and commercial fishers. In addition to the squid's reddish orange, blue and green colors spotted with chromatophores, or cells containing pigments which reflect light, the males acquire purple stripes running lengthwise on the underside of the mantle. These stripes, along with the visual cues produced by the chromatophores, are used in elaborate courtship displays.

Masters of the Language of Light

SHALLOW TO DEEP DARK WATERS

Think about living in a world of perpetual night with intense pressure pushing in on all sides. Survival here is a matter of being able to understand and process light signals. The light comes in different colours, some flashing, some faint. Certain light signals are used to attract mates. Some fish have light torches at the end of their forehead and others have chin poles to distract predators or lure prey. Still other fish live in a red world, giving off red light to communicate only to their own kind while remaining invisible to predators and potential prey who can't distinguish red light. And, of course, there is the opposite strategy for survival, staying dark, trying not to give off any signals at all while avoiding the illuminated gaze of predators.

More than any other ecosystem on Earth, the denizens of the mesopelagic, or middle waters, to the deeper waters of the world's oceans have evolved to use light as a tool. It's called bioluminescence, the biochemical emission of light, and humans are only beginning to understand this special language of light.

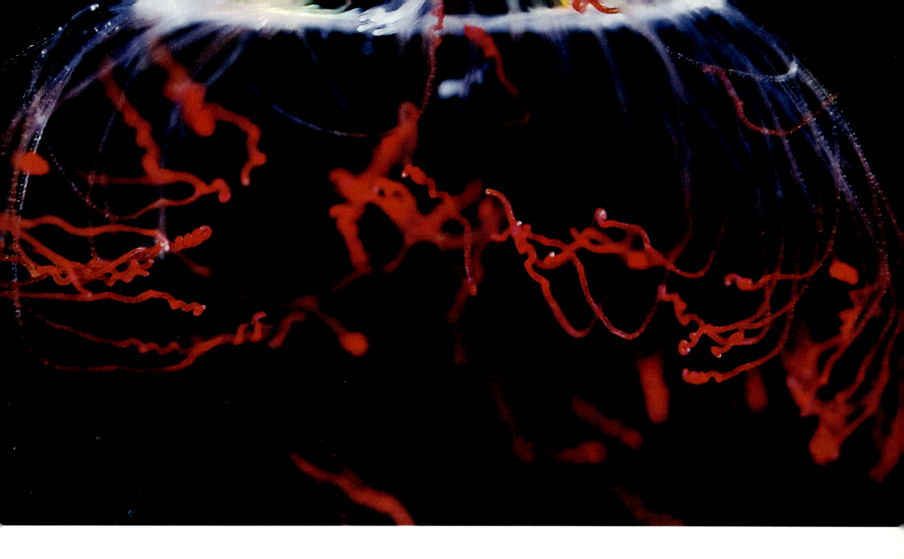

Photographer David Shale travelled with BBC Blue Planet and other expeditions to photograph the creatures that live in the deep sea. Along with others, he immersed himself in this cold, dark, high pressure world, watching and waiting for the flashing lights to go on and off, wondering what they mean. Photographers Solvin Zankl, Alexander Semenov and others brought their dedication and skills to expeditions in the Pacific, Indian, Atlantic and Arctic oceans.

There are fewer animals living in the middle to deep waters than in the surface waters and many of them are species that are new to science. Every expedition discovers suspected new species that may take years to categorize and name.

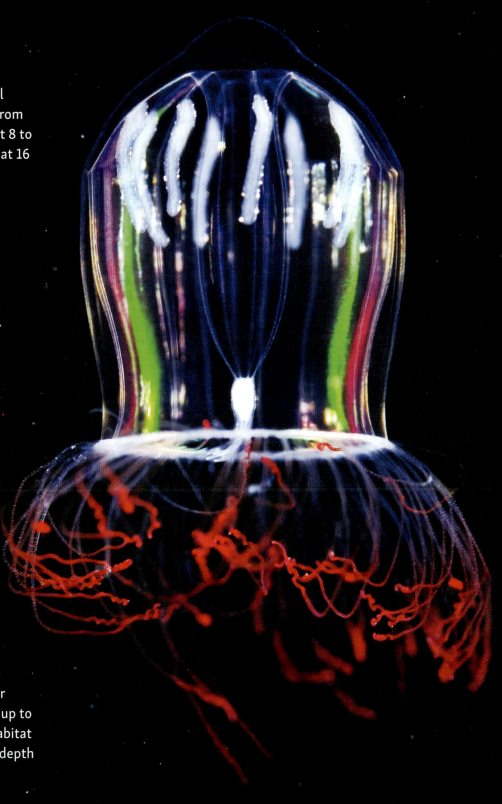

Crystal Jellyfish (left)
Aequorea, unidentified species

This crystal jellyfish was photographed in the Russian Sea of Okhotsk near the northern Kuril Islands. The hydrozoan crystal jellyfish glows from bioluminescence, especially when disturbed. At 8 to 16 inches (20 to 40 cm) long, they often cruise at 16 to 49 feet (5 to 15 m) below the surface.

Pink Helmet
Aglantha digitale

Photographed at a depth of 16 feet (5 m) below the surface at the White Sea Biological Station, near the Kandalaksha Nature Reserve in Russia, this tiny hydrozoan measured ⅜ inch (1 cm) in length. Pink helmets breed once a year during winter. They live for 1 year, occasionally up to 4 years in the Norwegian fjords. Pink helmet habitat extends across northern circumpolar seas at a depth of 165 feet to 2,300 feet (50 to 700 m).

Brown-banded Moon Jellyfish (left)
Aurelia limbata

This fast, muscular jellyfish, measuring 15 ¾ inches (40 cm) in diameter, was photographed near the Kamchatka shore in the western North Pacific. It lives at depths ranging from close to the surface to 3,300 feet (1,006 m) below the surface. Brown-banded moon jellyfish sometimes gather in dense groups of millions of individuals, devouring all organic material over a huge area. When in large groups they stay still in the water column, spreading their tentacles and oral lobes to feed.

Fried Egg Jellyfish or Egg-yolk Jellyfish
Phacellophora camtschatica

This big egg-yoke jellyfish was photographed in the Sea of Okhotsk, northern Kuril Islands, about 20 to 30 feet (6 to 9 m) below the surface. These jellyfish have a huge dome which can be up to 2 feet (60 cm) in diameter. They have sixteen clusters of a few dozen tentacles, which can be up to 20 feet (6 m) long. Their sting is so weak that jack fish often swim among the tentacles while crustaceans ride on the bell-shaped mantle, both stealing food, mostly other jellyfish and zooplankton that get caught in the sticky tentacles. Marine turtles eat the egg-yoke jellyfish and in their eagerness for a meal sometimes mistake plastic bags for the jellyfish and choke on the plastic.

Gulper Eel
Eurypharynx pelecanoides

The gulper eel is also called the pelican eel, the big-mouth gulper or the umbrellamouth gulper. This species' enormous mouth, long jaws and expanding stomach enable it to catch and eat large squid and deep-sea crustaceans. The gulper eel has dark velvety skin instead of scales and a long tail tipped with a pinkish light organ. Researchers speculate that this light organ may lure predators to attack the least vulnerable part of the gulper eel's body. The individual pictured here is 22 inches (56 cm) long but they can grow up to 2 ½ feet (75 cm) in length. Gulper eels live in all temperate and tropical seas, preferring depths from 500 to 6,000 feet (150 to 1,830 m).

Deepsea Batfish (left)

family Ogcocephalidae, unidentified species

Some species of batfish are deep-sea dwellers living on the ocean floor, but larvae can be found in the upper layers. This young one, slightly more than 1 inch (3 cm) long, was caught in a plankton haul in the Gulf of Mexico.

Bigeye Smooth-head

Bajacalifornia megalops

This 4 inch (10 cm) long fish inhabits the layer of water ranging from 2,600 to 4,900 feet (800 to 1,500 m) below the surface. This one was encountered on a research cruise on the *Henry Bigelow* to the Mid-Atlantic Ridge.

Dumbo Octopus
Stauroteuthis syrtensis

This little dumbo octopus is only 5 inches (12.5 cm) in diameter. It gets its name because of the two fins that stick out like ears and flap when it swims. The dumbo octopus in this photograph was found 2,700 feet (830 m) below the surface in the Gulf of Maine off the US east coast. It sometimes lives as much as 13,100 feet (4,000 m) below the surface but always stays a few hundred meters above the seafloor.

Spookfish or Barreleyes
Winteria telescopa

The 3 inch (7.5 cm) long spookfish is typically found at a depth between 1,640 to 2,300 feet (500 to 700 m) below the surface. Midwater to deep-sea fishes have sometimes evolved tubular eyes that are designed for looking up while swimming forward. The spookfish, however, has forward-looking tubular eyes. These normally restrict vision but with its accessory retinas to collect light it is able to detect movement from multiple directions.

Footballfish

Himantolophus paucifilosus

This 3 inch (7.5 cm) long footballfish may grow to nearly 6¼ inches (16 cm) in length. It lives mostly at a depth between 330 to 1,300 feet (100 to 400 m) below the surface in the Gulf of Mexico. Footballfish devour other small fish that are attracted to the light at the end of the flexible pole that sticks out from its forehead. The light is produced by bacteria that live on the tip of the pole.

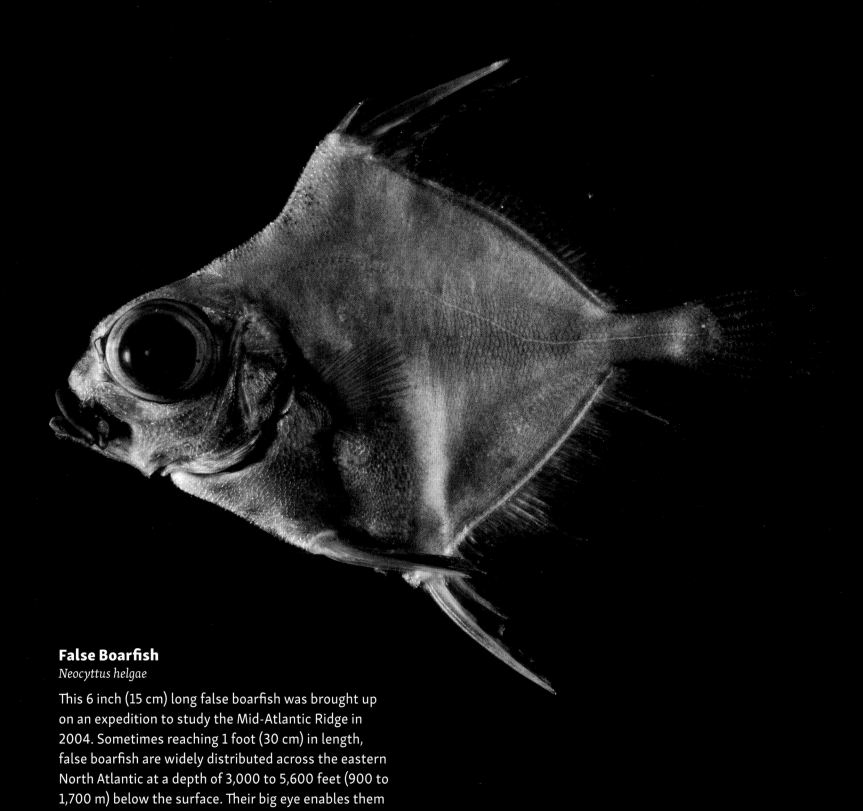

False Boarfish
Neocyttus helgae

This 6 inch (15 cm) long false boarfish was brought up on an expedition to study the Mid-Atlantic Ridge in 2004. Sometimes reaching 1 foot (30 cm) in length, false boarfish are widely distributed across the eastern North Atlantic at a depth of 3,000 to 5,600 feet (900 to 1,700 m) below the surface. Their big eye enables them to detect bioluminescent light flashes in their otherwise black world.

Common Fangtooth
Anoplogaster cornuta

This 4 inch (10 cm) long common fangtooth is a typical resident of the deep ocean. As it swims along in search of prey, the wide mouth and sharp inward curving teeth ensure that, once grasped, the crustaceans and other fish on which it feeds, can't escape. When the common fangtooth closes its mouth, its long lower teeth fit into sockets inside the upper lip.

Seadevil, Humpback Anglerfish or Deepsea Anglerfish (right)
Melanocetus johnsonii

Coming face to face with a seadevil up to 4,900 feet (1,500 m) below the surface, would be a Halloween nightmare for a fish of similar size. This 2 inch (5 cm) long seadevil was found above the Mid-Atlantic Ridge. These fish can reach 7 inches (18 cm) in length when mature. Males are much smaller than the females.

Deepsea Lizardfish
Bathysaurus ferox

Deepsea lizardfish rest on the bottom and feed on decapods and any other fish, even its own kind, that come close enough to catch. This one, about 12 inches (30 cm) in length, was collected and photographed on a research expedition to the Mid-Atlantic Ridge, north of the Azores, in 2004, in an area about 8,200 feet (2,500 m) below the surface. Mature lizardfish can reach a length of more than 2 feet (60 cm).

Frilled Shark (right)
Chlamydoselachus anguineus

A 3 foot (1 m) long frilled shark has its mouth open preparing to lunge at potential prey. The preferred prey, large squids and fish, get snagged on the rows of teeth before being swallowed whole. This species, which grows up to 6 ½ feet (2 m) in length, is called a living fossil due to its primitive features. It is rare to encounter one. It is thought to live between 66 and 4,900 feet (20–1,500 m), but typically occurs between 1,600 and 3,300 feet (500–1,000 m).

Ribbon Sawtail Fish or
Serpent Black Dragonfish
Idiacanthus fasciola

This fish, classified with the barbelled dragonfishes, is generally found between 1,640 feet (500 m) and 6,540 feet (2,000 m) beneath the surface in tropical and temperate waters of the Atlantic, Pacific and Indian oceans. The bioluminescent tip at the end of the pole, or barbel, lures fish close to investigate and allows the ribbon sawtail to seize a meal. Males range up to 1 9/16 inches (4 cm) and females up to 13 3/4 inches (35 cm) in length.

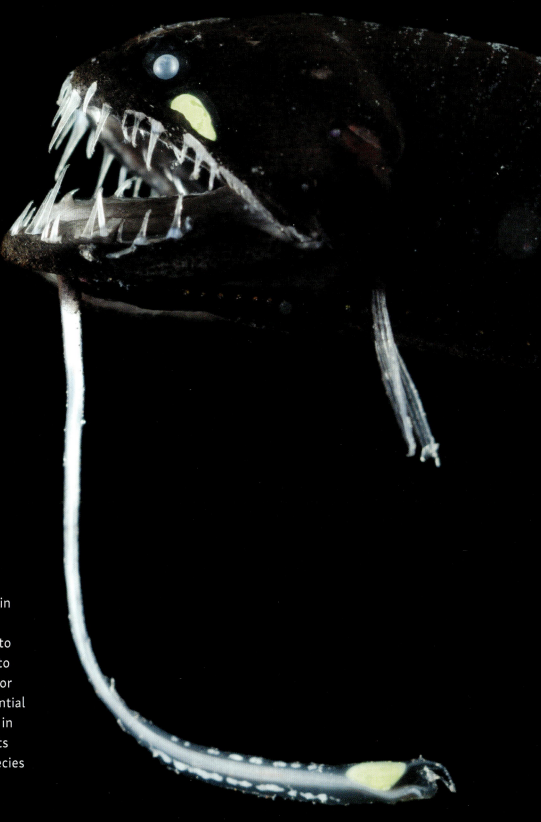

Black Dragonfish
Melanostomias biseriatus

This juvenile black dragonfish, 2 inches (5 cm) in length, wiggles like an eel through the middle to deep waters from 2,000 to 2,500 feet (620 to 760 m) below the surface of warm temperate to tropical waters. Note the bioluminescent lure or "fishing rod" for attracting and deceiving potential prey. The two cheek patch organs, which glow in the pitch black waters, are natural search lights used for hunting. Maximum length for this species is close to 10 inches (25 cm).

Deepsea Viperfish
Chauliodus sloani

This viperfish, 10 ⅝ inches (27 cm) long, shows off its chromatophores and bioluminescence. Living 3,280 feet (1,000 m) below the surface in global tropical to temperate waters, it has numerous light emitting organs that attract prey. The enormous teeth enable the viperfish to securely grab and hang on to its prey. This is essential as in the deep sea, prey encounters are rare and each chance for a meal has to be successful if the creature is to survive. However, sometimes a viperfish will bite into prey too large to be swallowed and the inward-bent teeth may make it impossible to let go. This might mean the end of both predator and prey.

Sea Butterfly
Limacina helicina

This mature sea butterfly, with a shell about ½ inch (1.3 cm) wide, looks like a snail waving big, dark earlike wings that appear to be growing out of its mouth. These wings evolved from the crawling foot of its gastropod ancestors. Sea butterflies live in open ocean Arctic waters where they sometimes comprise more than half of the zooplankton abundance. They serve as a main food for many marine species from sea angels to huge whales.

Sea Gooseberry
Euplokamis dunlapae

Not including the tentacles, which vary in length, the
sea gooseberry is ¾ inch (2 cm) long. The sea gooseberry
is a cold water comb jelly, or ctenophore, found in the
32 foot (less than 10 m) shallows of the White Sea in
Russia in April and May. Unlike other comb jellies, the
sea gooseberry is an athletic sprinter, and can even swim
backwards. The body of the gooseberry can be seen in
detail at left.

Dumbo Octopus
Stauroteuthis syrtensis

Also known as the glowing sucker octopus, this dumbo octopus was found 2,723 feet (830 m) below the surface. It is about the size of an orange. Unlike most octopuses whose suckers are adhesive, the suckers of the dumbo octopus have sensory cirri, or flexible tendrils. The cirri emit a faint blue-green bioluminescence presumed to attract prey which is then enveloped by the umbrella and eaten.

Hula Skirt Siphonophore
Physophora hydrostatica

The flamboyant hula skirt siphonophore is actually a colony that can reach 3 to 4 ¾ inches (8–12 cm) in length. The float, shaped like bells on top of each other, is the largest member of the colony. At the bottom of the float, a gas-emitting pore regulates the colony's buoyancy. The yellow, orange and purple tentacles, shaped like bananas, carry stinging cells. At the base of the float attached to threads are various small members of the colony with specialized tasks. The hula skirt siphonophore inhabits the deeper waters of the Atlantic, Pacific and Indian oceans.

Paper Argonaut or Paper Nautilus (left)
Argonauta species

This paper argonaut, ¾ inch (2 cm) in diameter, lives in the Gulf of Mexico. It has eight arms and no fins. The paper thin shell belongs to the female. Air is trapped in the shell to keep it buoyant and that is where she lays her eggs. Paper argonauts move around using jet-propulsion.

Cock-eyed Squid
Histioteuthis bonnellii

This cock-eyed squid, or umbrella squid, is 4 inches (10 cm) long but the species can grow up to about 1 foot (30 cm) long. Found deep in the mesopelagic at around 656 feet (200 m) below the surface, the cock-eyed squid has one large eye and one small eye, one looking up and one down, along with a complex arrangement of photophores to counterbalance the difference in light coming from above and below.

The Bottom Dwellers

THE CONTINENTAL SHELF TO THE ABYSSAL PLAIN

If you were asked by a creature from another planet what is the most common geographical feature of the planet Earth, you would have to say it is the vast abyssal plain at the bottom of the ocean. Yet only less than 1 percent of all the abyssal plain in the ocean has been explored and surveyed and we have almost no idea of the diversity of life to be found there.

At the bottom of the sea, sea cucumbers stand tall, waving their tentacles. A no-eyed animal, the sea cucumber is essentially a mouth, a gut and an anus. It has stumps for legs that allow it to stand or walk on the bottom and occasionally push itself up to catch sediment flowing in a current that, with any luck, might be nutritious. Other bottom dwellers, like anemones and sea lilies, are fixed to the bottom. Certain worms, fish and crustaceans bury themselves in the sediment while others, such as squid, octopus and jelly species, cruise just above the bottom picking out prey and able to jet away to avoid predation. More than anything, the bottom dwellers rely on the perpetual rain of everything from larval to whale-sized dead bodies

that come from the waters above them. A whale dying and sinking to the bottom can support an entire ecosystem for hundreds of creatures. Living whales, dolphins and other marine mammals provide regular supplies of poo — a bonanza of nutrients for developing animals that not only increases the productivity of marine ecosystems but also helps remove thousands of tons of carbon from the atmosphere, contributing to climate stabilization.

Alexander Semenov, who took many of these photographs, loves to dive in the very cold waters of Russia, from the White Sea in the Arctic to the Okhotsk Sea in the Russian Far East, uncovering surprising diversity in relatively untouched seas.

bring the food to the sea cucumber's mouth. Each sea cucumber is about 6 inches (15 cm) across. This forest-like underwater scene recalls a passage from *The Lord of the Rings* with the sea cucumbers looking like miniature Ents preparing for battle.

¹ Emperor Shrimp
Zenopontonia rex (formerly *Periclimenes imperator*)

² Sea Cucumber
unidentified species

Some 50 feet (15 m) below the surface on the bottom of the Bohol Sea in the Philippines, a ½ inch (1.5 cm) long emperor shrimp uses a sea cucumber as a fortress, mobile home and feeding station. The shrimp can extend its claws down to the seabed to feed by sifting the sandy bottom as well as grabbing and munching on cucumber poo. Scientists have not determined if the sea cucumber gains anything from the relationship but the emperor shrimp may help discourage predators.

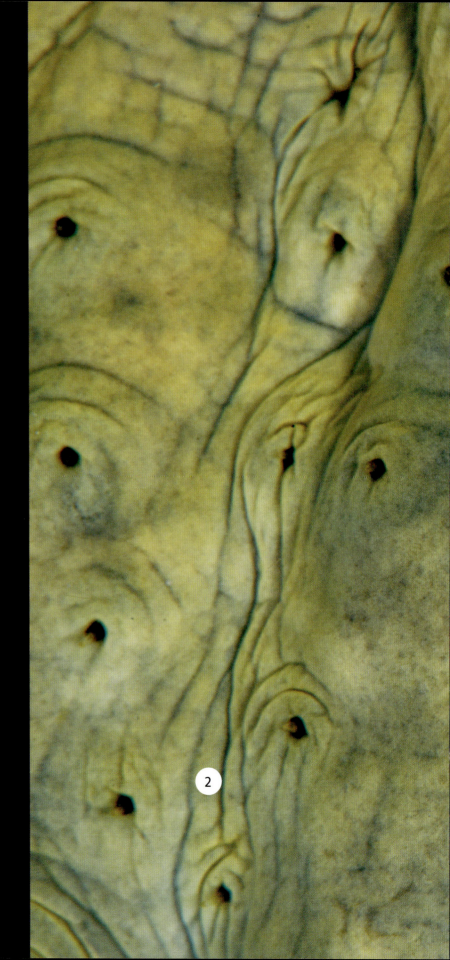

Sea Apple
Pseudocolochirus violaceus

Typically found on coral reefs, sea apples are colorful, rounded sea cucumbers with tube-like feet standing about 8 to 12 inches (20 to 30 cm) tall. Their tentacles, extending from a crown, wave in the water column to catch food particles. One by one, the tentacles deliver food to the mouth. Like many other sea cucumbers, they can expel their internal organs to confuse or distract the predator, or even squirt a toxin into the water when threatened by nosy fish or a hungry squid or octopus, then slowly walk away. This sea apple was found off Rinca Island in Komodo National Park, Indonesia, at a depth of 35 to 70 feet (10 to 20 m) below the surface but they can live much deeper.

Sea Grapes
Molgula griffithsii

Sea grapes, a kind of sea squirt, are marine invertebrate filter feeders with a tough outer "tunic" or sac. They are about 1⁹⁄₁₆ inches (4 cm) in diameter. This species is shown in a group, which is rare as they are not colonial. These cold-water sea grapes were photographed between 39 and 49 feet (12–15 m) below the surface in the White Sea. Sea grapes are translucent with two protruding siphons. They are found subtidally, attached to slow-moving submerged objects or organisms.

¹ Sea Slug
Chlamylla intermedia

² Sea Slug
Flabellina nobilis

³ Oaten Pipes Hydroid
Tubularia indivisa

In this photograph taken 80 to 100 feet (25 to 30 m) below sea level in the White Sea of Russia, two sea slugs ⅜ inch (1 cm) and 1⁹⁄₁₆ inch (4 cm) long feed on opposite sides of the stem of an oaten pipes hydroid, a preferred diet item. Hydroids, a life stage for most animals of the class Hydrozoa, are small predators related to jellyfish. Besides the oaten pipes hydroid, these sea slugs feed on various plankton, detritus and other invertebrates on the ocean bottom.

Sea Slug
Flabellina verrucosa

This shell-less marine sea slug, less than 1 inch (2 cm) long, was found near the bottom of the White Sea in Russia. Sea slugs often have bright colors to deter predators. Like all species in the class of gastropods, they have tiny razor-sharp teeth. Most sea slugs have two pairs of tentacles on their head that are used primarily for smell as well as a small eye at the base of each tentacle. The protruding structures on their backs, called cerata, act as gills. This sea slug uses its ability to incorporate stinging nematocytes, stolen from its jellyfish and cnidarian prey, which are kept in little white sacs on the end of its cerata as a defence against certain fish and seastar predators.

Deep Water Stalked Jellyfish
Lucernaria bathyphila

This stalked jellyfish, measuring 1¼ inch (3 cm), lives 82 feet (25 m) below sea level. It is one of the few jellyfish found in the deep waters of the White Sea. The species name, "bathyphila," means lover of the deep. Their soft bodies are fragile and they live in waters protected from currents or water disturbance.

Feather Duster Worm
Pseudopotamilla reniformis

The feather duster worm, or sabellid polychaete, builds a tube from its own fluid strengthened by sand and bits of shell. This one, about 2 ½ inches (6.5 cm) long was found 39 feet (12 m) deep on the bottom of the White Sea. Feather duster worms have a crown of feeding appendages in fan-shaped clusters that project from their home-made tubes. The protruding feathery branchiae function as gills.

Sea Anemone (right)
Stomphia coccinea

Widely distributed in the cool to cold waters of the northern hemisphere, this sea anemone lives about 66 feet (20 m) below the surface in the White Sea. It attaches itself to rocks and shells and catches plankton floating in the water using its tentacles which can number up to 80. If attacked by predator starfish or a large sea slug, it can detach itself and swim away, re-attaching somewhere else. The species name, coccinea, means scarlet and refers to the anemone's often reddish to orange-striped color. This individual is at the maximum size for the species, 3 to 4 inches (8–10 cm) long.

Lemon Paddle Worm
Phyllodoce citrina

The Swedish common name for this species translates as "lemon paddle legs" and is inspired by the several hundred yellowish flattened cirri, or appendages, found along their back side. Lemon paddle worms have tiny dark red eyes. This 10 inch (25 cm) long individual can be found in shell gravel about 33 to 49 feet (10 to 15 m) below the surface in the Russian White Sea. Their habitat extends to muddy bottoms up to 886 feet (270 m) below the surface.

King Ragworm (right)
Alitta virens

This king ragworm, a kind of annelid worm related to land-based earthworms, lives in the White Sea and measures about one foot (30 cm) long. A mature ragworm can reach 20 inches (50 cm) in length. King ragworms burrow into the sand or sediment, or hide under stones at about 33 to 49 feet (10 to 15 m) below the surface. They emerge to snatch small worms, mollusks and crustaceans, striking snake-like with their two teeth and bringing the prey back into the safety of their burrow. They also scavenge for whatever is available and even eat algae. King ragworms recognize chemical signals from other members of their species which are thought to serve as a warning that active predators are nearby and that they should stay in their burrow.

Sea Cucumber
Peniagone diaphana

The sea cucumber pictured here is about 3 inches (7.5 cm) long. It was found on the ECOMAR expedition using a remote operated vehicle (ROV). It was swimming about 3¼ feet (1 m) above the seafloor at a depth of 8,500 feet (2,600 m) near the Charlie Gibbs Fracture Zone of the Mid-Atlantic Ridge. The ROV cameras observed the sea cucumber drifting with its head down. ROVs enable scientists to glimpse what these gelatinous deep-sea creatures look like in their habitat. Most sea cucumbers and jellies disintegrate if collected through net sampling and brought to the surface due to the change in temperature from 3°–4°C on the sea floor to up to 20°C or more at the ocean's surface.

Sea Cucumber
Peniagone porcella

This 3 ⅛ inches long (8 cm) sea cucumber lives just above the bottom of the Charlie Gibbs Fracture zone on the Mid-Atlantic Ridge at a depth of about 8,500 feet (2,600 m) below the surface.

Stalked Barnacles

Neolepas, unidentified species

These 3 inch (7.5 cm) long stalked barnacles were found 9,200 feet (2.8 km) below the surface in the Longqi (Dragon's Breath) Vent Field on the South West Indian Ridge in the Indian Ocean. The barnacles live on active hot vents called black smokers that are up to 20 feet (6 m) high. Black smokers are geothermal vents on the seabed that eject superheated water and sulphide minerals.

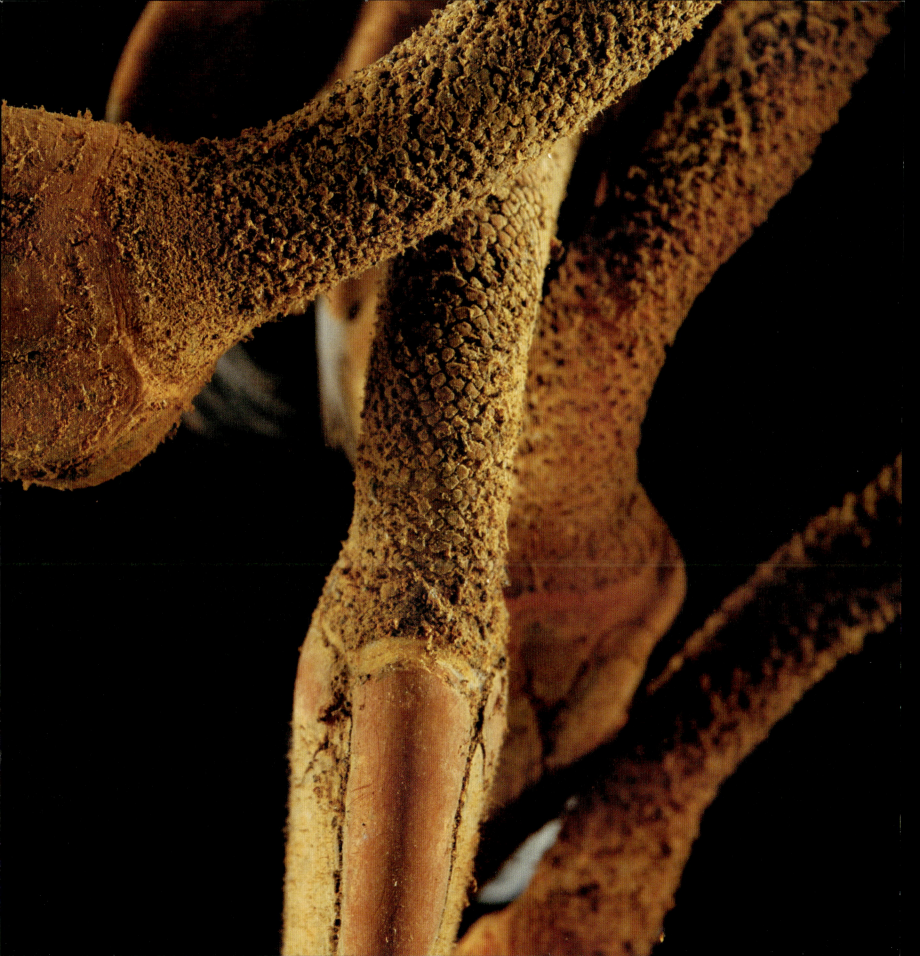

Amphipod

Paramphithoe hystrix

Like many creatures that live on the bottom of the ocean, this ¾ inch (2 cm) long amphipod has no common name. Amphipods are an order of crustaceans, shrimp-like in form, that live up to 30,000 feet (9,100 m) below the surface. This one was found in Arctic waters of the Norwegian Barents Sea.

North Atlantic or Spoonarm Octopus
Bathypolypus arcticus

Usually found near the bottom between 700 and 2,000 feet (200–600 m) below the surface, this octopus is about 2¾ inches (7 cm) long and feeds mainly on brittle stars. The surface of the mantle and arms are notably "warty." The female broods her eggs for more than 400 days during which time she stops eating and slowly wastes away as she metabolises her own body to provide energy for her to stay alive while guarding the eggs and young. This is one of the longest brooding periods for any octopus.

Decapod Crustacean
Eiconaxius gololobovi

Decapods, so named for their ten feet, or appendages, are an order of ghostly white deep-sea crustaceans related to crabs, lobsters and shrimp. This male decapod is about ¼ of an inch (0.8 cm) long and was discovered on a coral seamount on an expedition to explore the Southwest Indian Ridge, 2,300–4,300 feet (700–1,300 m) below the Indian Ocean. It was formally described as a new species in 2018.

¹ Stenopods
Spongiocaris tuerkayi

² Venus Flower Basket Glass Sponge
probably *Euplectella aspergillum*

This male stenopod (a type of shrimp) is a little more than ¼ inch (.8 cm) long and the female is ½ inch (1.4 cm) long. They live and breed inside this glass sponge which was found 2,430 feet (740 m) below the surface on the bottom of the South West Indian Ridge in the Indian Ocean. This stenopod species was described in 2016.

Porcupine Crab
Neolithodes grimaldii

The porcupine crab, a species of king crab, is covered from head to foot in long spines and has a leg span that can reach 40 inches (1 m) across. This one was collected from the Mid-Atlantic Ridge. The species is widely distributed on the bottom of the sea along the continental slope and abyssal plain as well as on deep-sea ridges in both the western and eastern North Atlantic. It has been recorded at depths of up to 17,159 feet (5,230 m) below the surface.

Bristle or Polychaete Worm
unidentified species

This unnamed deep-sea bristle worm, about 2 inches (5 cm) long, was found on the Mid-Atlantic Ridge during a research cruise on the American research ship *Henry Bigelow*. Note the muscular mouthparts around the jaws and the sensory tentacles.

Benthic Hydrozoan
Ptychogastria polaris

This jelly, found only 16 feet (5 m) below the surface off Tasiilaq, East Greenland, often lives near the bottom on a type of kelp, *Saccharina latissima*. The jelly was first described following an 1875 voyage to the polar sea aboard the H.M.S. *Alert* and *Discovery*. This animal is adult size, close to ¾ inch (1.5 to 2 cm) in length. It swims in the water like other jellies, moving its tentacles; sometimes it stays on the bottom

Giant Deepsea Isopod (right)
Bathynomus giganteus

Deepsea isopods are the marine equivalent of the woodlouse, their land based relative. In the deepsea, however, they are an example of deep-sea gigantism. Compared to most isopods, giant deepsea isopods are much larger. This specimen is nearly 8 inches (20 cm) long, but there is a "supergiant" isopod species that can grow up to a record 30 inches (76 cm) in length and weigh 3.7 pounds (1.7 kg). These isopods are deep-sea scavengers with large compound eyes and two pairs of antennae.

White-plumed Anemone or Giant Plumose Anemone

Metridium farcimen

Standing more than 40 inches (1 m) tall, this huge anemone lives along most of the North Pacific Rim from Alaska south to Catalina Island, California, and from the Commander Islands off Kamchatka to Sakhalin and the Kuril Islands north of Japan. This one was photographed at a depth of about 54 feet (16.5 m) beneath the surface in the northern Kuril Islands. White-plumed anemones are known to be long-lived, some thriving for more than 100 years.

Squat Lobster
probably *Munida* or *Galathea* species

A resident of the coral communities on seamounts found on the deep Southwest Indian Ridge, this 3 inch (7.5 cm) long female squat lobster carries her bright red eggs under her tail. Egg-carrying females are said to be "gravid." Many new species of squat lobsters have been identified in the Indian Ocean in recent decades.

Deepsea Basket Star (right)
Gorgonocephalus lamarckii

This deepsea basket star, less than 2 inches (5 cm) across, was found in the North Atlantic off Iceland. It has five arms that branch out dividing repeatedly to create branchlets. To feed, this basket star perches in an elevated position on the bottom 500 to 3,300 feet (150 to 1,000 m) deep, extending its arms in a basket-like fashion. The branches and branchlets twist and coil, capturing small crustaceans that approach too close. The arms are covered in tiny hooks to grasp the prey. Along with the tube feet, these arms bring food to the mouth, located on the underside of the central disc.

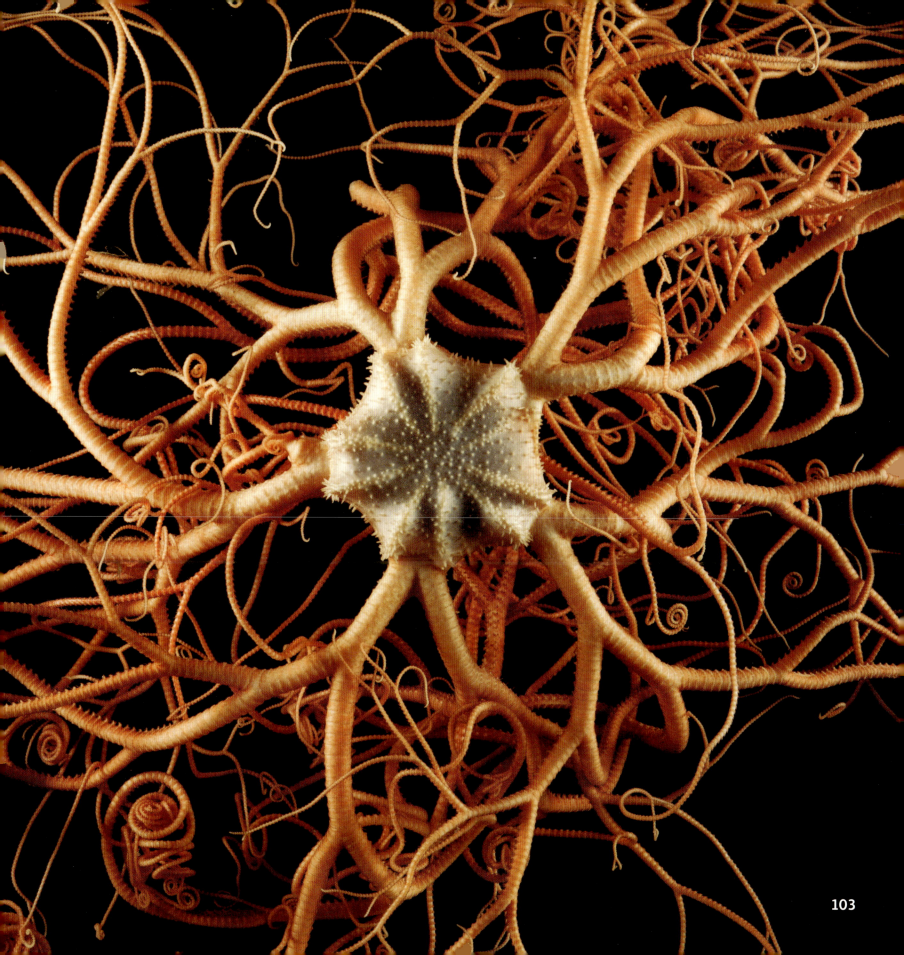

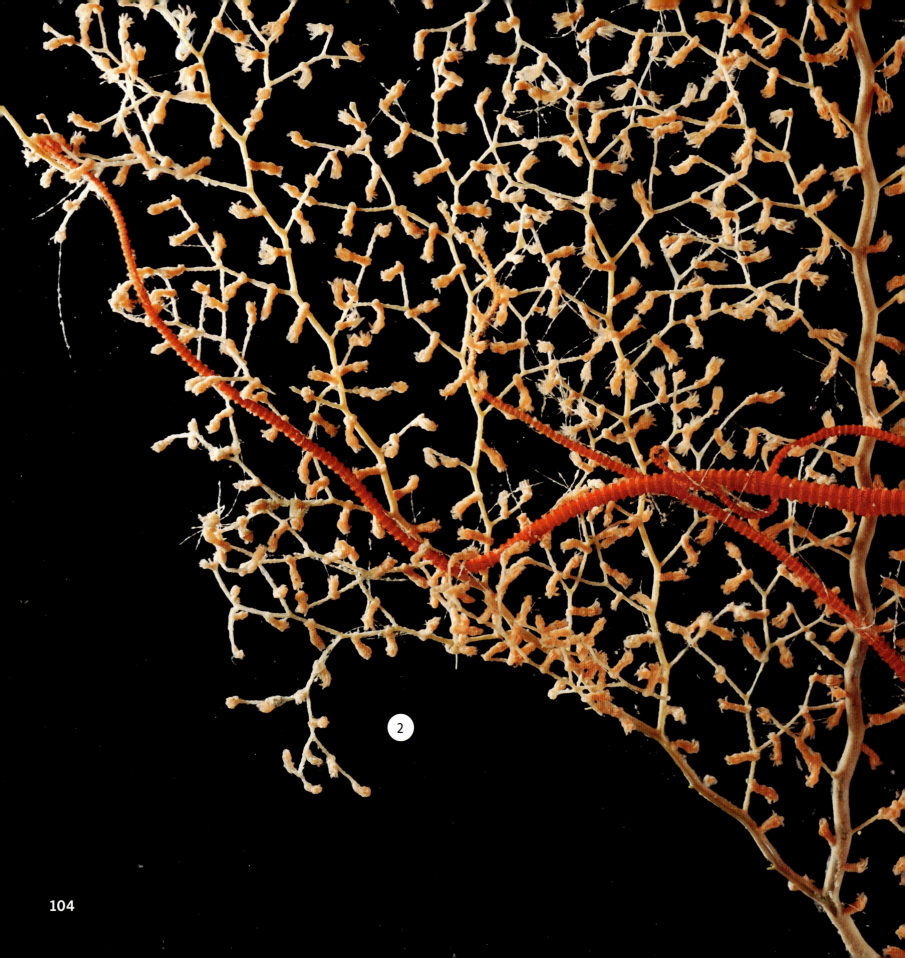

¹ Brittle Star
unidentified species

² Gorgonian Sea Fan Coral
unidentified species

Deep in the North Atlantic, on the Mid-Atlantic Ridge, a brittle star, 4 inches (10 cm) across, straddles gorgonian sea fan coral, 6 to 8 inches (15 to 20 cm) long. Both are filter feeders. The brittle star takes advantage of the hard coral to gain a firm footing against the current, allowing it to seize any food particles drifting by. Brittle stars are also scavengers and detritivores that sometimes consume small worms or crustaceans as well as decomposing plant and animal parts and feces.

¹ **Gray Whale**
Eschrichtius robustus

² **Eelpout**
family Zoarcidae; unidentified species

³ **Boneworms or Zombie Worms**
Osedax, unidentified species

⁴ **Deepsea Octopus**
unidentified species

⁵ **Crab**
unidentified species

In October 2019, the ROV *Nautilus* probed the seafloor 10,620 feet (3,238 m) below the surface of Davidson Seamount in Monterey Bay National Marine Sanctuary, off California. They came upon a fresh whale fall—a newborn 13 to 16½ foot (4 to 5 m) long gray whale that had recently died and sunk to the bottom and was now providing food for various deep-sea species large and small. Scavengers like eelpouts stripped the skeleton of blubber, while bone-eating *Osedax* worms were tunnelling into the skeleton to suck lipids (fats) from the bones. The most visibly active were the dozens of ghostly white deep-sea octopuses drawn to the carcass.

REFLECTIONS FROM THE SURFACE

Who are these creatures of the sea that seem so alien to us? They make us stop and wonder how such beings can possibly exist on our planet. Yet they also have the power to inspire scientists, doctors, children, poets — everyone — to rethink our place in the great scheme of life and about the necessity to change our ways.

The photo on the opposite page shows how, even in the ocean, nature wastes nothing. Each organism in the food chain supports and makes possible the whole and, eventually, even the largest becomes food for the smallest. Humans , too, are in the food chain, but our value to nature leaves much room for improvement. The days are numbered in which humans can freely exploit the oceans, evading their responsibilities to planet Earth and disrupting the lives of the strange and wonderful creatures that live in ocean.

We need a massive clean-up, cool-down and a rethink about how we treat the ocean and its precious residents and the homes, or ecosystems, where they live. We cannot continue overfishing and catching hundreds of thousands of unintended whales, dolphins, seabirds, turtles, sharks and rays. We cannot keep drilling into the seabed for hydrocarbon and minerals. We cannot continue to dispose in the sea our industrial equipment, garbage, plastics, sewage, agricultural chemicals and waste water.

We have long had a love affair with the beaches and nearshore waters of the sea but we haven't paid attention to what is really going on down below. The 2020s and 2030s may be the last decades before all-out exploitation and development, along with climate change, turn the ocean into an empty polluted pond. It is up to all of us to make changes in our own lives and to elect leaders to implement the society-wide changes leading to a future for life in the sea as well as for the Earth itself.

GLOSSARY

Annelid worm: The large phylum, or group, of segmented worms including ragworms, earthworms and leeches.

Bell: The umbrella-shaped hood or dome which forms the main part of the body of a jellyfish from which tentacles are suspended.

Benthic: Referring to the ecological region at the bottom of the sea.

Bioluminescence: The production and emission of light by a living organism including vertebrates, invertebrates, fungi and bacteria.

Cerata: Protrusions on the body of a nudibranch.

Ceriantharia: Subclass of tube-dwelling anemones.

Chromatophore: Cell or group of cells containing pigments which reflect light.

Cirri: Flexible tendrils.

Cnidarian: Any invertebrate marine animal (jellyfish, sea anemone, coral) belonging to the phylum Cnidaria characterized by stinging structures in the tentacles surrounding the mouth.

Dinoflagellate: One-celled aquatic organism that swims using two flagella, a major part of the plankton and a common bioluminescent organism.

Dorsal fin: The fin on the back of most fish, whales and dolphins and other aquatic vertebrates.

Ephyra: Free-swimming larval stage of jellyfish development.

Gastropod: A mollusk such as a slug or snail, sometimes with a shell and a distinct head carrying sensory organs.

Hemichordate: A marine invertebrate of the phylum Hemichordata of wormlike marine animals such as an acorn worm.

Heteropod: A group of pelagic gastropods (mollusks).

Hydroid: A life stage for animals related to jellyfish in the class Hydrozoa. Most hydroids are colonial with individual polyps specialized for feeding and reproduction while others are solitary.

Hydromedusa: The medusa form of a hydrozoan produced asexually by budding from a hydroid.

Hydrozoan: A class of small predatory mostly marine animals, some solitary and some living in groups. The colonies of the colonial species can be large and, in some cases, the specialized individual animals cannot survive outside the colony. Hydrozoans are related to jellyfish and corals and are sometimes referred to as jellyfish.

Mantle: The umbrella-shaped top of most cnidarians including jellyfish, also sometimes referred to as the bell; also used to describe the enclosed area between the tentacles of cephalopods, squid and octopods.

Medusa: One of the two main body types of cnidarians besides polyps. The medusa form is shaped like a bell or umbrella.

Nematocysts: The stinging cell in the tentacles of cnidarians such as jellyfish. A nematocyst is a capsule containing a barbed tube that produces a sting that can be toxic to predators and prey.

Nudibranch: A shell-less soft-bodied marine snail or gastropod with appendages on the back and sides used for respiration.

Octopod: Any of the eight-armed animals in the order Octopoda of cephalopod mollusks including argonaut and octopus species.

Pectoral fin: The pair of side fins, sometimes called breast fins or flippers.

Pelagic: Living in the water column and not on the bottom of the sea.

Photophore: The luminous spot on various fish and cephalopods, a glandular organ specialized to produce light (see bioluminescence).

Phyllosoma: The larval stage of spiny, slipper and coral lobsters.

Phytoplankton: Plant plankton (see plankton).

Plankton: Organisms that drift through the water, contrasted to those that swim. A primary food for many animals, plankton includes both phytoplankton, plant-based organisms including microscopic algae, and zooplankton, animal organisms such as copepods and many larval stages of fishes and invertebrates.

Polychaete: An annelid worm, characterized by a segmented body; bristle worm.

Polyp: The hollow, columnar, sessile (attached) form of Cnidarians (as opposed to the medusa form).

Pteropod: Sea butterfly, free-swimming marine sea snail or sea slug.

Siphonophore: One of the swimming or floating colonies of marine hydrozoans in the order Siphonophora such as the Portuguese man o' war.

Tunicate: A marine invertebrate member of the subphylum Tunicata. It is part of the Chordata, a phylum which includes all animals with dorsal nerve cords and notochords. Their name derives from their unique outer covering or "tunic," which is formed from proteins and carbohydrates, and acts as an exoskeleton.

Ventral: The underside of an animal.

Zooplankton: Animal plankton (see plankton).

INDEX